Quantum Universe

Author Muriel Fernandes
All rights reserved to the author
Registered in the national library

Index

1. Introduction
   1.1 Context of Quantum Physics
   1.2 Importance of Consciousness
2. Intersection of Quantum Physics and Spirituality
   2.1 Fundamental Concepts
   2.2 Spiritual Connections

3. The Interconnection of the Cosmos
    3.1 Principle of Non-Locality
    3.2 Social Responsibility

4. The Power of Consciousness
    4.1 Influence of Observation
    4.2 Intention and Reality

5. Personal and Collective Transformation
    5.1 Challenges to Materialist Paradigms
    5.2 Integration of Spiritual Perspectives

6. Practices that Cultivate Consciousness
    6.1 Meditation and Visualization
    6.2 Mindfulness

7. A Call to Action
    7.1 Personal Reflections
    7.2 Agents of Change

8. The Future of Human Consciousness
    8.1 Cultural Shifts
    8.2 Domino Effects

9. The Potential of Quantum Consciousness in Human Evolution
    9.1 Quantum Psychology
    9.2 Quantum Spirituality

10. Creativity and Innovative Thinking

   10.1 Superposition and Creativity

   10.2 Instant Insights

11. Ethics and Global Responsibility

   11.1 Ethical Choices

   11.2 Quantum Ethics

12. Conclusion: Embracing the Evolution of Quantum Consciousness

   12.1 Interconnection between Quantum Physics and Consciousness

   12.2 The Role of Creativity and Responsibility

13. The Quantum Realm Experienced by Me

   13.1 Personal Experiences

   13.2 Reflections on the Journey

14. The Alchemist of the Future

   14.1 Theory of the Seven Keys

   14.2 Contributions to the Understanding of the Interconnection between Mind, Universe, and Spirituality

Introduction

How many times do we find ourselves immersed in questions about the true nature of the universe and the meaning of our existence? These questions have echoed throughout history, provoking curious minds. Since ancient times, religion, philosophy, and science have tried to unravel these mysteries, but often their answers do not encompass the complexity of life.

It is in this fascinating scenario that Quantum Physics stands out, illuminating unknown paths to understanding reality. With its astonishing discoveries, this discipline challenges our deeply held beliefs and invites us to a profound reflection on the nature of the universe and human consciousness. The intriguing behavior of subatomic particles inspires us to expand our horizons and explore new perspectives that may guide us on our evolutionary journey.

In this book, I invite you to embark on a unique journey where Quantum Physics intertwines with spirituality. We will explore everything from fundamental theories to their practical applications, revealing the implications for our understanding of consciousness. My desire is for these pages to become a source of

inspiration, providing insights that help you deepen your understanding of the nature of the universe and your own existence. Together, we can open the doors to a new era of evolution and transformation.

Chapter 1

In this chapter, I invite you to explore the perspective of Quantum Physics on the nature of reality and its intersection with the philosophy of science. We will discover how quantum theory provides us with a new way of understanding the world, revealing that subjectivity is an intrinsic aspect of physical reality. This understanding leads us to rethink our approach to science and the world around us, with significant implications for our interaction with life.

Traditionally, the classical view of reality is based on concepts such as time, space, and causality, where the physical world exists independently of human observation and can be described objectively. In this perspective, causality establishes a clear relationship between cause and effect.

However, Quantum Physics challenges this view by introducing the idea that subatomic particles can exist simultaneously in multiple states. Before being observed, these particles have no defined position or velocity; they possess a superposition of possibilities. Only when observed do they collapse into a specific state, suggesting that reality is shaped by observation.

Another fascinating concept brought forth by Quantum Physics is quantum entanglement, which proposes that particles can be instantaneously connected, even when separated by great distances. This connection challenges classical notions of space and causality, indicating a new interpretation of reality. Furthermore, measurement in Quantum Physics affects the results of experiments, highlighting that scientific objectivity has its limits, as observation involves a subjective element. The famous double-slit experiment perfectly illustrates this idea, where the mere observation of a particle's position alters its trajectory.

The quantum field theory also brings forth the notion that particles can emerge from the quantum vacuum, transforming the classical conception that matter is the foundation of reality. This theory suggests that the quantum vacuum is an active entity, generating particles constantly through quantum fluctuations.

These fundamental principles of Quantum Physics not only question traditional concepts such as space, time, and causality but also open doors to a new interpretation of reality. This new perspective has profound implications for the philosophy of science and can facilitate the integration of different forms of knowledge, including spirituality.

Quantum mechanics represents a revolution in our understanding of the universe, challenging traditional notions, especially the idea of causality. While classical physics provides us with an orderly and deterministic view of reality—where every effect results directly from a cause—quantum mechanics introduces us to a world where indeterminacy and probability become fundamental factors. This new perspective is not only fascinating; it forces us to reevaluate what we know about the nature of the real.

A central concept that illustrates this change is indeterminacy. In classical physics, if we know all the initial conditions of a system, we can accurately predict its future behavior. At the subatomic level, quantum mechanics tells us that we cannot simultaneously know a particle's position and velocity with absolute precision.

Heisenberg's Uncertainty Principle reveals that we are dealing with a range of possible outcomes, each laden with uncertainty. This idea challenges the deterministic logic that underpins the classical view of causality.

Quantum superposition makes us rethink causality. Imagine that a particle can exist in multiple states at the same time until it is measured. This concept is exemplified in the double-slit experiment, where particles like electrons create interference patterns that can only be understood if we consider them entangled in a superposition of states.

When measurement occurs, the particle "collapses" into a specific state, leading us to question how causes manifest in a world where observation plays a crucial role. Quantum entanglement further challenges the classical notion of causality. When two particles are entangled, measuring one of them instantaneously affects the other, regardless of the distance between them. This inexplicable connection suggests that causality does not follow the rules we know. If information can be transmitted instantaneously, what does this mean for our understanding of space and time?

Quantum mechanics shows us that observation is not a neutral act. The simple act of measuring a system alters its state, implying that the outcome of an experiment is not just a consequence of initial conditions but also of the act of observation. This dynamic reveals that causality is not linear and predictable; the consciousness and action of the observer play an active role in what manifests.

These revelations from quantum mechanics broaden our understanding of how events unfold and invite us to explore the interconnectedness of all phenomena in the universe. As we move away from the certainties of classical physics, we are challenged to embrace the complexity and beauty of the quantum world, where indeterminacy and interdependence are the new rules of the game.

As we conclude this first chapter, it is evident that quantum mechanics is not just a collection of equations and scientific theories; it is, above all, an invitation to deep reflection on the nature of reality and our place within it. The journey we have begun reveals a world where indeterminacy, probability, and interconnectedness become the protagonists in a narrative that challenges the certainties cultivated throughout history.

The classical view of reality, grounded in deterministic logic, offered us an understanding of the world based on clear relationships between cause and effect.

However, as we enter the quantum realm, we encounter a reality where predictability dissolves into a dance of possibilities. Heisenberg's Uncertainty Principle teaches us that, at the subatomic level, we cannot simultaneously know a particle's position and velocity with precision. This forces us to reevaluate our assumptions about what is possible to know. Uncertainty, far from being an obstacle, becomes an intrinsic characteristic of nature, revealing a universe more complex and fascinating than we could imagine.

The concept of quantum superposition, which shows us that a particle can exist in multiple states simultaneously, challenges our intuitions about reality. The double-slit experiment exemplifies this complexity.

The idea that a particle can behave like a wave, interfering with itself, leads us to question what it truly means to "exist." Here, observation is not neutral; rather, it influences the outcome, showing that we are co-creators of the reality we experience. This understanding leads us to reflect on the role we play in our lives and how our perceptions shape the reality around us.

The phenomenon of quantum entanglement presents us with a new view of the interconnectedness between particles and, by extension, between all of us. The idea that two particles can remain instantaneously connected, regardless of distance, suggests that the boundaries we establish between ourselves and the world are more fluid than they seem. This interdependence invites us to reconsider our relationships, both on a personal and collective level. In a world where everything is interconnected, the awareness of our actions and thoughts becomes even more crucial.

On our journey, we encounter the question of causality. Quantum mechanics challenges the classical notion that every effect must have a direct and predictable cause. Indeterminacy and chance become central, leading us to ponder what it truly means for something to "cause" another thing. This new way of thinking teaches us to accept that not everything can be explained logically and linearly, encouraging us to be more flexible in our approaches.

As we advance in our exploration of quantum mechanics, we are introduced to different interpretations that challenge us to rethink what we know about consciousness, time, and space. The many-worlds interpretation, for example, suggests that each quantum event results in branching universes, creating parallel realities. This idea not only destabilizes our perception of causality but also leads us to consider the multiplicity of experiences that exist in potential.

What becomes clear in this chapter is that quantum mechanics is not merely a scientific curiosity; it is a transformative approach that offers us new ways to understand reality. It challenges us to open our minds and hearts to the complexity and beauty of the universe. As we move away from the certainties of classical physics, we are invited to embrace uncertainty as an intrinsic part of existence. This acceptance not only enriches our understanding of the world but also helps us connect more deeply with each other and with the cosmos.

As we continue our journey, I invite you to remain open to these new ideas and consider how they may impact your understanding of the world and your everyday life. Quantum physics teaches us that, in an interconnected and dynamic universe, every thought, action, and choice has the potential to resonate beyond what we can see. This is an invitation for each of us to become active agents in creating our own reality, recognizing that we are part of a vast and ever-evolving whole.

The next chapter will take us even deeper into this dialogue, exploring how these quantum concepts can be applied to the evolution of human consciousness. Together, we will uncover the possibilities that arise when science and spirituality intertwine, illuminating our path toward a deeper understanding of the nature of existence and our role within it.

Chapter 2: The Quantum View of Reality

In this chapter, we will delve into fundamental concepts of Quantum Physics, such as Heisenberg's uncertainty principle and the double-slit experiment, which challenge the classical view of reality. These elements guide us toward understanding wave-particle duality, a central theme that helps us rethink the nature of matter and energy.

Heisenberg's uncertainty principle teaches us that the precision in measuring the position and velocity of a particle is limited. The more precisely we try to determine one of these properties, the less precision we will have in measuring the other. This relationship reflects the probabilistic nature of Quantum Physics, questioning the classical notion of na objective and deterministic reality.

The double-slit experiment illustrates this duality. In it, individual particles seem to pass through both slits at the same time, creating na interference pattern on the detection screen, as if they behave like waves. This revelation challenges the classical view that particles must follow a specific path. As we examine these concepts, we begin to understand how Quantum Physics opens new doors to thinking about reality.

As we progress, we will see that Quantum Physics transforms our understanding of the physical world and invites us to reflect on the connection between science and consciousness. This relationship brings significant changes to our perception of ourselves and the universe around us.

One of the main differences from the classical view lies in the understanding of time and space. In the classical perspective, these elements are seen as separate and objective entities, existing independently of the observer. Causality is considered linear, where each cause generates a predictable effect. The idea of freedom and free will is often regarded as na illusion, as everything appears to be governed by immutable natural laws.

In contrast, Quantum Physics reveals that time and space are interconnected and subjective, shaped by observation. Causality ceases to be linear and becomes a complex web of possibilities influenced by the presence of the observer. This new perspective offers us a renewed understanding of free will, freedom, and creativity. Instead of seeing ourselves as products of natural laws, we are invited to recognize the active role we play in creating reality

This quantum view leads us to rethink fundamental concepts such as free will and creativity. Freedom is no longer na illusion but na essential part of our existence, highlighting our ability to shape the world around us. Creativity is redefined as the collective capacity to create new possibilities from the interconnections between things. By challenging the classical view, Quantum Physics invites us to explore our relationship with the world in a deeper way.

Continuing our journey, we will delve into the link between Quantum Physics and consciousness, as well as its applications in spiritual healing practices. This intersection between science and spirituality has been debated for centuries. While science seeks to understand nature through observation, spirituality explores reality through inner experience. Quantum Physics offers na opportunity to overcome this dichotomy, creating common ground between these fields.

By revealing the complexity of reality and the interconnectedness of the physical world, Quantum Physics finds parallels with many spiritual traditions that emphasize the fundamental unity of the universe. This perspective challenges the mechanistic view of classical science, which sees the universe as a machine composed of isolated parts. Instead, Quantum Physics presents us with a dynamic and interconnected system, where each part influences the other. This new view has the potential to expand our understanding of consciousness, mind, and spirituality.

Quantum Physics helps us integrate science and spirituality, recognizing that both are essential for a deeper understanding of reality. This integration allows us to explore new ways of understanding both ourselves and the world around us.

Furthermore, Quantum Physics can clarify phenomena considered paranormal or spiritual. The idea that consciousness is linked to the quantum world may offer explanations for experiences such as telepathy and precognition. Concepts like energy healing can be seen as non-local influences of consciousness on reality. These ideas, often traditional in spirituality, may find support in Quantum Physics, providing a basis for reconciliation between science and spirituality.

As we deepen our understanding of the interaction between consciousness and the quantum world, we have the opportunity to explore spiritual and paranormal aspects more profoundly. This perspective transforms our view of reality and can have significant ethical and social implications.

Quantum Physics teaches us that all actions are interconnected and that our choices impact the entire reality. This understanding motivates us to adopt a collaborative and responsible approach to the environment and society. Instead of seeing ourselves as isolated individuals, Quantum Physics reminds us of our connection and the importance of acting in a conscious and sustainable manner.

This new vision highlights the importance of collaboration and mutual respect. Understanding our interconnectedness encourages us to work together rather than focusing solely on individual interests. This approach can help overcome social conflicts and promote a more united and compassionate coexistence.

The double-slit experiment is fascinating, as it illustrates wave-particle duality, a central concept of quantum mechanics. This experiment reveals how subatomic particles, such as electrons and photons, can exhibit behaviors of both particles and waves, depending on the conditions of observation.

First, let's understand the setup of the experiment. In it, a source of particles is directed toward a barrier with two parallel slits. Behind the barrier, there is a detection screen that records where the particles hit.

From the classical perspective, we would expect that, when passing through the slits, the particles would follow a determined path, creating two distinct peaks on the screen corresponding to each slit.

However, the reality is surprising. When we conduct the experiment, we observe na interference pattern on the screen, even when sending particles one at a time. This means that the particles do not behave as simple particles, but rather as waves. When the waves pass through both slits, they interfere with each other—some add together, creating areas of reinforcement, while others cancel out, resulting in areas of darkness. The result is a pattern of alternating fringes, characteristic of wave interference.

The most intriguing aspect of the experiment is that, when we try to observe which slit the particle is passing through by placing a detector at one of the slits, the interference pattern disappears. What was once a wave-like behavior transforms into typical behavior of classical particles, generating two impact regions corresponding to each slit. This leads us to reflect: the simple act of observing alters the behavior of the particle, as if it decides to be a particle instead of a wave the moment we observe it.

What does all this teach us about duality? At the quantum level, particles do not fit into the classical categories of "particle" or "wave." They have a dual nature that allows them to behave like waves under certain conditions and like particles in others. This duality is a fundamental characteristic of quantum reality and challenges us to rethink our intuitions about what it means to be a particle or a wave.

Observation plays a fundamental role in the double-slit experiment, significantly altering the behavior of the particles and challenging our intuitive understanding of reality. When we send particles, such as electrons or photons, one at a time toward the barrier with two slits, we observe that, if no measuring device is present, these particles behave like waves, creating na interference pattern on the detection screen. This pattern is characteristic of waves, where the waves passing through both slits overlap, generating areas of reinforcement and cancellation.

Now, here comes the intriguing part: when we place a detector at one of the slits to observe which path the particle is taking, the interference pattern disappears. Instead, what we see on the screen resembles what we would expect from classical particles, where they seem to pass through a specific slit. This leads us to reflect on the profound implication that the simple act of measuring or observing a particle alters its behavior.

This phenomenon is explained by the concept of "wave function collapse." In quantum mechanics, before we measure, a particle exists in a superposition of states—that is, it can be in multiple positions or paths at the same time. When we make the measurement, this superposition "collapses" into a defined state. By observing the particle, we force a choice among the possibilities, causing it to behave like a classical particle rather than a wave.

The implications of this are profound, leading us to question the nature of reality and the role of the observer. This suggests that reality is not a fixed fact, but something more fluid and interdependent.

The double-slit experiment illustrates na essential aspect of quantum mechanics: the effect of the observer. The presence of a measuring device not only provides us with information about the particle but also alters the very nature of what we are measuring. This challenges our traditional notions of objectivity and reality, suggesting that the interaction between the observer and the observed system is a fundamental part of the quantum experience.

Quantum mechanics challenges our traditional notions of reality and objectivity in fascinating ways. Wave-particle duality reveals that particles, such as electrons and photons, can behave both as particles and as waves, depending on the conditions of observation. This duality forces us to rethink the idea that objects have a fixed nature. Reality becomes more complex and multifaceted, showing that the way we observe a phenomenon can influence its behavior.

Another important concept is Heisenberg's Uncertainty Principle. It teaches us that we cannot simultaneously know the position and velocity of a particle with absolute precision. The more precisely we try to measure one of these properties, the less precision we will have in measuring the other. This relationship of uncertainty suggests that reality is not as objective as we thought; instead, it is probabilistic, challenging the idea that we can have complete knowledge of what is happening around us.

When we talk about wave function collapse, we enter another intriguing aspect. In quantum mechanics, before a measurement, a particle exists in a superposition of states. When we make a measurement, this superposition "collapses" into a defined state. This implies that reality is not something fixed until it is observed. The act of observing is not a passive recording; it shapes reality itself.

Moreover, interconnectedness and non-locality are fundamental concepts. The phenomenon of quantum entanglement shows us that particles can be instantaneously connected, regardless of the distance separating them. When two particles are entangled, measuring one of them instantaneously affects the other, no matter the distance. This interconnectedness challenges the classical view that objects are isolated entities, suggesting that reality is a dynamic and interdependent network.

Quantum mechanics emphasizes the importance of the observer in defining reality. Unlike the classical view, which considers reality to exist independently of observation, quantum mechanics suggests that the observer's consciousness plays na active role in creating what we consider real. This idea raises profound philosophical questions about the nature of perception and consciousness, challenging the notion that we can be objective observers.

Finally, quantum mechanics introduces the idea that many phenomena cannot be predicted with certainty, but only with probabilities. This contrasts with the classical view, where future events can be determined based on initial conditions. The probabilistic nature of quantum mechanics suggests that the universe is intrinsically random at fundamental levels, leading us to reconsider the idea of absolute determinism.

As we reach the end of this chapter, it is essential to reflect on the implications of quantum mechanics for our understanding of the world and ourselves. What we have explored here is not just a series of complex physics concepts; it is na opportunity to reevaluate how we perceive reality, objectivity, and our own existence.

Quantum mechanics offers us a new lens through which we can observe the universe. By challenging the classical view that reality is something fixed, it invites us to consider a dynamic and interconnected world. This shift in perspective is na exciting and challenging journey, showing us that reality is not merely a backdrop where things happen, but a space where interactions, choices, and observations shape what we consider real.

When we talk about wave-particle duality, we explore the idea that everything in life has multiple facets. Just as particles can behave like waves, we too are multifaceted beings. Our identities, experiences, and perceptions change with the context and interactions we have with the world. This awareness encourages us to be flexible and to appreciate the nuances of each situation.

Heisenberg's uncertainty principle reminds us that we cannot have absolute knowledge of everything. This uncertainty is not something to be feared; it is na intrinsic part of the human experience. Accepting that not everything can be predicted or controlled frees us from the limitations we impose on ourselves. Life is a flow of possibilities, and the best experiences often arise from the unexpected. This openness to uncertainty makes us

more resilient and creative.
Wave function collapse presents us with the idea that reality may be shaped by observation. This leads us to question our role in the world. Are we mere spectators or active participants in creating our reality? Quantum mechanics suggests that we are co-creators of what we experience. Our choices and perceptions influence the world around us. This responsibility can be intimidating, but it is also empowering. By recognizing our active role, we can act more consciously, aligning our decisions with our values.
The interconnectedness among particles, evidenced by quantum entanglement, reminds us of our own interdependence. We live in a world where everything is connected—our actions reverberate not only in our lives but also in the lives of others and the environment around us. This understanding invites us to cultivate a more compassionate and collaborative approach in our interactions. By recognizing that we are part of a greater whole, we are encouraged to act with responsibility and empathy, seeking collective well-being.
Furthermore, quantum mechanics challenges us to rethink our idea of objectivity. The fact that observation can alter the behavior of particles leads us to question what it truly means to "see" or "know." In a world where our perceptions are shaped by experiences and contexts, it becomes clear that truth can be as complex a construct as reality itself.

This awareness helps us to be more tolerant and open to others' perspectives, understanding that each person brings a unique narrative that deserves to be heard.

As we delve deeper into these concepts, it is crucial to remember that quantum mechanics is not just a scientific theory; it is a philosophy of life that invites us to explore the complexity of existence. It challenges us to step out of our comfort zones, to question what we know, and to open our minds to new possibilities. This journey of discovery can be exhilarating and disconcerting, but it is also na opportunity for growth and transformation.

Chapter 3 – The Quantum Atom

The quantum atom is the foundation of Quantum Physics, consisting of a central nucleus that contains protons and neutrons, while electrons orbit around this nucleus. According to Quantum Physics, electrons occupy specific quantized energy levels, a phenomenon known as energy quantization. One of the first atomic models to apply quantum concepts was proposed by Niels Bohr in 1913, who suggested that electrons circulate in fixed orbits, or "energy levels," instead of following continuous movements. This idea helped explain the quantized nature of the energy of electrons.

However, the Bohr model did not adequately describe the wave nature of electrons nor the impossibility of simultaneously determining their position and momentum, as stated by Heisenberg's Uncertainty Principle. The contemporary view of the atom is supported by quantum mechanics, which allows us to understand the wave nature of electrons and their interactions with the nucleus. The wave function mathematically represents the probability of finding na electron at a given point in space. Additionally, the Pauli Exclusion Principle establishes that two electrons cannot share the same set of quantum numbers, limiting the number of electrons in each energy level and providing a better understanding of the structure of atoms and molecules.

The study of the quantum atom deepens our understanding of the physical world and finds crucial applications in fields such as chemistry, materials physics, and semiconductor technology. Throughout history, various atomic models have been proposed to explain the structure and behavior of the atom. Thomson's model, from 1897, conceived the atom as a positive sphere with electrons embedded within it. Although innovative, this model failed to explain the energy distribution of electrons. In 1911, Rutherford's model described the atom as a planetary system, but it still lacked complete explanations of energy distribution.

Only with the advent of Quantum Physics did atomic models begin to make sense. Bohr's model, proposed in 1913, was pioneering in explaining the energy distribution of electrons, stating that they occupy well-defined orbits around the nucleus. When na electron transitions between these orbits, it emits or absorbs a quantum of energy. Other quantum models emerged, such as Schrödinger's, which used the wave equation to describe the behavior of electrons, and Dirac's, which incorporated the theory of relativity.

Quantum Physics clarified phenomena such as energy quantization and the Pauli Exclusion

Principle. Quantization arises from the restriction of electrons to defined energy levels, and the Pauli Exclusion Principle stipulates that two electrons cannot simultaneously share the same quantum state. This quantization is one of the pillars of Quantum Physics that challenges the classical view, introducing the idea that energy is quantized, meaning it takes on discrete values. Max Planck introduced this concept in 1900, proposing that energy is emitted or absorbed in packets called quanta.

In quantum theory, the energy of na atom is quantized and determined by discrete values. This is due to the wave-particle duality of matter, which indicates that matter can manifest as either a particle or a wave. When na electron is confined in na atom, it assumes the behavior of a standing wave, being restricted to certain wavelengths, resulting in discrete energy values.

The quantization of energy is fundamental to explaining quantum phenomena, such as the emission of light by excited atoms, which occurs only at discrete wavelengths. Furthermore, quantization is linked to Heisenberg's Uncertainty Principle, which states that it is impossible to simultaneously measure the position and velocity of a particle with absolute precision. This happens because measuring the position interferes with the velocity, generating uncertainties. In summary, the quantization of energy is a fundamental concept of Quantum Physics that challenges the classical view of reality.

The Pauli Exclusion Principle – A Foundation of Quantum Physics

The Pauli Exclusion Principle, one of the pillars of Quantum Physics and Chemistry, was proposed by Austrian physicist Wolfgang Pauli in 1925. This principle establishes that two electrons within na atom cannot share the same quantum numbers, meaning they cannot have the same values of spin, angular momentum, and energy.

This rule is vital for unraveling atomic structure and molecular formation, explaining why atoms have electronic shells, filling the inner shells before the outer ones.

Moreover, the principle is crucial for understanding the stability of atoms and why some are more stable than others. The implications of the Pauli Exclusion Principle extend beyond Chemistry and Quantum Physics, reaching other disciplines such as solid-state theory. This is evident in understanding why electrons in metallic materials form a continuous energy band, as opposed to discrete values. This observation is central to discerning the electrical and magnetic properties of materials. In summary, the Pauli Exclusion Principle is a cornerstone of Quantum Physics and Chemistry, providing insights into the electronic organization around the atomic nucleus and being crucial for molecular formation.

The Practical Contribution of Quantum Physics

Quantum Physics has vast practical applications in various areas, and technology associated with the atom plays a significant role. The development of technologies based on quantum principles has driven remarkable advances in electronics, computing, and medicine. One concrete example is laser technology, devices that emit coherent and highly directed light. These devices have wide-ranging applications, from optical communication to medicine, and their operation is rooted in quantum principles, such as stimulated emission, influenced by the atomic structure of materials.

Another crucial application of Quantum Physics is nuclear magnetic resonance (NMR), a medical technique that generates images of the inside of the human body. NMR provides accurate and non-invasive diagnoses for various medical conditions. Its principle is based on the behavior of atomic nuclei in a magnetic field, and its interpretation involves quantum concepts such as spin and nuclear relaxation.

In addition to these examples, there are numerous other applications of Quantum Physics, such as spectroscopy, quantum cryptography, and semiconductor technology. These applications have become viable thanks to a deep understanding of atomic structure and associated quantum phenomena.

Iconic Experiments in Quantum Physics

Numerous experiments have been conducted throughout the history of Quantum Physics to confirm theoretical predictions and expand our understanding of the quantum nature of the world. Let's highlight some renowned experiments that have contributed to solidifying quantum theory.

The double-slit experiment is one of the most famous. In this experiment, electrons or photons are directed through two slits, generating na interference pattern on the detection screen. This phenomenon clearly evidences the wave nature of matter and how interference occurs even with individual particles, illustrating wave-particle duality.

Another fundamental experiment is the Stern-Gerlach experiment, which involves sending atoms through a non-uniform magnetic field, allowing for the observation of the deviation of their trajectories. This experiment helped demonstrate the existence of electron spin and its relation to the magnetic properties of atoms.

The Franck-Hertz experiment, conducted in 1914, was crucial for confirming quantum theory and understanding atomic structure. In this experiment, electrons are sent through a mercury tube, measuring their kinetic energies before and after colliding with mercury atoms. The results corroborated the existence of discrete energy levels in atoms and energy quantization.

These and other experiments reinforced quantum theory and expanded our understanding of quantum nature. Additionally, their practical implications are significant in various areas, including laser technology, nuclear magnetic resonance, and other quantum innovations.

The Main Difference Between Bohr and Rutherford Atomic Models

The main difference between Bohr and Rutherford atomic models lies in how each describes the structure and behavior of electrons in relation to the atomic nucleus.

Rutherford Model: Proposed in 1911, it presents the atom as a planetary system. In this model, the atomic nucleus, containing protons and neutrons, is extremely small and dense, while electrons orbit around it, much like planets around the sun. Although this idea was revolutionary in introducing the concept of a central nucleus, the model failed to explain how electrons remain stable in their orbits. According to classical physics, na electron in continuous circular motion should lose energy and spiral into the nucleus, something that does not happen in reality.

Bohr Model: The Bohr model, presented by Niels Bohr in 1913, improved Rutherford's view by introducing the quantization of electron orbits. Bohr suggested that electrons do not orbit at any distance around the nucleus, but rather in fixed and defined orbits, known as "energy levels." These levels are quantized, allowing electrons to occupy only certain specific orbits. When transitioning between these orbits, they emit or absorb quanta of energy. This model better explains the stability of atoms and the emission of light at discrete wavelengths, but it still does not fully address the wave nature of electrons.

The Rutherford model describes the position of electrons in relation to the nucleus in a simple yet innovative way for its time. The concept of the atom as a mini solar system, with electrons orbiting around a dense nucleus, was based on alpha particle scattering experiments in which Rutherford observed that most particles passed directly through the atom, but some were deflected at large angles, indicating the presence of a dense nucleus.

Although this model represented a significant advancement, it failed to adequately explain the stability of electrons in their orbits. This limitation led to the development of more advanced models, such as that of Niels Bohr, which introduced the idea of quantized orbits for electrons.

As we conclude our exploration of the quantum atom, it is important to reflect on the impact of this knowledge on our understanding of the world and our daily lives. What we discussed in this chapter is not just a series of scientific theories; it is a fascinating journey that leads us to rethink the nature of reality and our own existence.

When we look at Rutherford's atomic model, we see how it revolutionized our understanding by introducing the idea of a central nucleus, where most of the atomic mass resides. However, this representation also reminds us of the limitations of scientific knowledge. Rutherford opened the door for new explorations but could not explain how electrons remained stable in their orbits. This uncertainty leads us to reflect on how much there is still to discover and how science is in constant evolution.

With the advent of the Bohr model, we entered a new phase of understanding. Bohr introduced the idea that electrons occupy quantized energy levels. This notion of quantization is fundamental, not only to explain atomic structure but also to make us think about the nature of energy and reality. When we consider that electrons can jump between these orbits, emitting or absorbing quanta of energy, we realize that reality is not a continuous sequence of events but a series of discrete transitions.

Just like electrons, our lives can also be marked by moments of change and transformation.

The quantization of energy and the Pauli Exclusion Principle help us understand the structure of the atom and the complexity of the matter around us. The Pauli Exclusion Principle, which states that two electrons cannot occupy the same quantum state, is a rule that applies to many aspects of our lives. It teaches us about the importance of individuality and diversity. Just as each electron needs its own space and state, each of us has a unique role to play in the world.

Considering the practical applications of Quantum Physics, such as laser technologies and nuclear magnetic resonance, it is fascinating to realize how these principles translate into innovations that impact our daily lives. Lasers have transformed communication, medicine, and the entertainment industry, while magnetic resonance provides us with non-invasive diagnostics that save lives. These technologies are a testament to the power of science and human capability to turn knowledge into progress.

The experiments that solidified quantum theory, such as the double-slit experiment and the Stern-Gerlach experiment, show us that science is a continuous quest for answers. Each experiment is a piece of the puzzle, and each discovery leads us to new questions. The wave-particle duality, evidenced by the double-slit experiment, reminds us that reality is often more complex than it seems. Just as electrons can behave as either particles or waves, we too can play different roles in our lives, adapting to circumstances and interactions.

These quantum concepts are not just scientific abstractions; they have profound implications for our understanding of life and the universe. Quantum mechanics teaches us that uncertainty is na intrinsic part of reality. Heisenberg's Uncertainty Principle reminds us that we cannot always have all the answers, and that is acceptable. In a world full of complexity and change, learning to live with uncertainty can make us more resilient. Just like electrons in superposition, we have the potential to explore multiple possibilities in our lives.

As we reflect on the quantum atom and its implications, we are invited to rethink our own existence and the role we play in the universe. Just as electrons interact with the nucleus and with each other, we are interconnected in a complex web of relationships. Our actions have repercussions that extend beyond ourselves, impacting the world around us. This interconnectedness invites us to act with responsibility and compassion, recognizing that every choice we make can resonate in many directions.

The journey towards understanding the quantum atom reminds us of the importance of curiosity and the pursuit of knowledge. Science is not na isolated discipline; it is a collective effort involving collaboration and the exchange of ideas. Just as the scientists who came before us, we have the responsibility to continue this quest. We must question, explore, and never cease to marvel at the mysteries of the universe.

At the end of this chapter, I invite you to reflect on how the concepts discussed here can apply to your own life. How can you incorporate the flexibility and adaptability of electrons into your daily interactions? What space can you create to accept uncertainty and let it drive you toward new discoveries? How can you become na active agent in creating your reality, shaping it through your choices and intentions?

Quantum Physics offers us a new perspective on life, challenging us to consider the complexity and beauty of existence. As we continue to explore these concepts, may we be inspired to live with greater awareness, curiosity, and empathy. The universe is vast and full of possibilities, and each of us is na essential part of this great mystery. May this journey of discovery continue to guide us towards a deeper understanding of ourselves and the world around us.

Chapter 4: The Quantum Electron

The quantum electron, na elementary particle, orbits the atomic nucleus and is fundamental to the chemical properties of elements. According to quantum theory, the electron is not a classical particle with defined position and velocity, but a wave of probability described by a wave function. This function contains information about the energy, position, and momentum of the electron. The wave nature of the electron is central to quantum theory, leading to intriguing phenomena such as tunneling, which allows the electron to "cross" energy barriers that, at first glance, seem insurmountable.

Moreover, quantum theory predicts states in which the electron can coexist in multiple locations, a concept known as quantum superposition. Studying the quantum electron is essential for understanding atomic and molecular structure, with applications in electronics, medicine, and quantum computing.

Spin and Angular Momentum

Two fundamental properties of electrons in Quantum Physics are spin and angular momentum, both of which are quantized, meaning they take on discrete values. Angular momentum measures rotation around na axis, while spin refers to the intrinsic rotation of the electron. Angular momentum is particularly relevant because it is linked to the energy of the electron.

As a Hermitian operator, it commutes with the Hamiltonian, being solutions to the Schrödinger equation for different angular momenta.

Electron Spin

Spin is a property unique to the electron, which has no classical analogy and is exclusive to quantum theory. It is intrinsic and inherent to the electron, unable to be explained by observable characteristics such as charge or mass. The implications of spin are vast: it affects the fine structure of atomic spectral lines, influences the formation of chemical bonds, and is vital in technologies such as nuclear magnetic resonance, widely used in medical diagnostics and scientific research.

Tunneling Effect

The tunneling effect is a fascinating quantum phenomenon in which particles can cross energy barriers even without sufficient energy, challenging the laws of classical physics. This behavior arises from the wave nature of matter, which behaves as a probability wave. In classical mechanics, a particle would not be able to cross na energy barrier without the necessary energy. However, in quantum mechanics, the particle is represented by a wave function that describes the probability of finding it in different positions and energy states. Thus, there is a chance that the particle can tunnel through the barrier, even though classical physics would advise against such a possibility.

The tunneling effect has relevance across various disciplines, encompassing particle physics, quantum electronics, and chemistry. In particle physics, this phenomenon explains the radioactive decay of unstable nuclei. In quantum electronics, it is applied in devices such as scanning tunneling microscopes and potential barrier tunnels. In chemistry, the tunneling effect elucidates molecular reactions on solid surfaces, such as the hydrogen reaction in catalysts. The ability of a particle to cross na energy barrier, defying classical logic, reflects the quantum nature where particles behave in unpredictable ways.

The Electron in Electric and Magnetic Fields

The behavior of the electron in electric and magnetic fields is essential in Quantum Physics. In these fields, the electron experiences forces that can be described by the Schrödinger equation, fundamental in quantum mechanics. Under a uniform electric field, na electron is accelerated, gaining kinetic energy and potentially transitioning between energy states, emitting or absorbing electromagnetic radiation. In a uniform magnetic field, the electron is deflected and may rotate perpendicular to the field vector.

Interactions in electric and magnetic fields have implications in various areas, such as nuclear magnetic resonance, medicine, and electric motors. The quantum properties of the electron are crucial for technological advancement, especially in creating electronic devices such as transistors and microchips. These transistors, which form the basis of electronic circuits, and microchips, which contain billions of transistors, are fundamental in modern technology and drive advances in various fields.

Quantum Physics is essential for the development of advanced electronic devices, allowing for the control of the quantum properties of electrons. Understanding the behavior of these particles influences sectors such as electronics, medicine, and data processing.

Valence Electrons

Valence electrons play a fundamental role in the formation of chemical bonds, being essential for understanding how atoms interact and combine to form molecules. Let's explore some points that highlight the importance of these electrons:

1. **Determination of Reactivity:** Valence electrons are the outermost electrons of na atom and are directly involved in chemical interactions. The configuration of these electrons determines the reactivity of the element. For example, atoms with one or two valence electrons, such as alkali metals, tend to easily lose these electrons. In contrast, elements with seven valence electrons, such as halogens, typically seek to gain electrons to complete their valence shell. This dynamic is crucial for understanding how different elements react with each other.

2. **Formation of Chemical Bonds:**
   - **Covalent Bonds:** In this type of bond, atoms share valence electrons to achieve a more stable electronic configuration. For example, in the water molecule (H_2O), oxygen shares its valence electrons with the valence electrons of hydrogen, forming covalent bonds that hold the atoms together.
   - **Ionic Bonds:** Here, one atom donates one or more valence electrons to another atom, resulting in the formation of ions, which can be cations or anions. This transfer of electrons leads to electrostatic attraction between oppositely charged ions. A classic example is sodium chloride (NaCl), which forms when na electron is transferred from sodium (Na) to chlorine (Cl), resulting in Na^+ and Cl^- ions.

   2. **Molecular Stability:** The way in which valence electrons are distributed and their total number influences the stability of the formed molecules. Atoms strive for a stable electronic configuration, and the interaction between valence electrons helps create durable molecules.

As we reach the end of this journey through the intriguing world of the quantum electron, I invite you to reflect on the beauty and complexity of the nature that surrounds us. What we have learned about the electron, its properties, and its fundamental role in the structure of matter is a window into better understanding the universe we live in.

Imagine the subtle dance of electrons around the atomic nucleus. This movement is not just a matter of physics; it is the essence of what forms everything around us—from the water we drink to the air we breathe. Each interaction and each chemical bond shapes matter and life as we know it, connecting us intimately to the wonders of science and nature.

As we delve deeper into quantum physics, we are challenged to rethink our perceptions of reality. The tunneling effect, superposition, and electron spin show us that the universe is much more complex and surprising than our intuitions suggest. This is na important lesson: life, like quantum physics, is full of uncertainties and possibilities. Sometimes, we are led to cross barriers that seem insurmountable, and it is in this quest for understanding that we find growth and transformation.

Moreover, science teaches us about the interconnectedness of everything. Each electron, each atom, each molecule has a role to play in the vast tapestry of existence. Just as valence electrons influence chemical reactions, we too have the power to influence the world around us, whether in our daily interactions or in our life choices.

Chapter 5: The Quantum World

In this chapter, we will delve into the fascinating quantum world, one of the fundamental pillars of Quantum Physics. This domain, radically different from the classical world we know, is essential for understanding quantum phenomena such as superposition and entanglement, which will be discussed in subsequent chapters.

Quantum Physics has revolutionized our perception of the universe, revealing that things are not exactly as they seem. In this peculiar world, particles can exist simultaneously in multiple states, lacking a well-defined position or velocity until they are measured. This uniqueness is part of what makes Quantum Physics so intriguing and challenging.

We will explore some key properties of the quantum world, such as wave-particle duality, the uncertainty principle, and wave function collapse. Additionally, we will discuss how these characteristics relate to practical applications of Quantum Physics, including quantum computing technology.

A central concept we will also address is String Theory, which seeks to unify all physical laws into a comprehensive theory. This theory suggests that fundamental particles are not dimensionless points, but rather extremely small vibrating strings (on the order of 10^{-33} centimeters). String Theory offers an explanation for the fundamental forces of nature, such as gravity and the electromagnetic, weak, and strong forces, within a single theoretical framework. It also predicts the existence of additional dimensions beyond the three spatial and one temporal dimension, which may help elucidate phenomena still considered inexplicable by conventional Physics.

Despite its potential as a unifying theory, String Theory still lacks direct experimental evidence. However, it has generated intense research and debate within the scientific community. Many theoretical physicists believe that this theory could be the key to unraveling some of the greatest mysteries of the universe.

Relativistic quantum mechanics, in turn, combines quantum mechanics with Einstein's theory of special relativity. This integration is vital for describing phenomena that occur at high energies or at speeds close to the speed of light, where classical physics no longer applies. In this context, particles are governed by the Dirac equation, which accounts for relativistic energy and momentum. Additionally, relativistic quantum mechanics postulates the existence of antiparticles, which possess properties analogous to normal particles but with opposite charges. This approach has enriched our understanding of elementary particles and the functioning of particle accelerators, such as the LHC (Large Hadron Collider), also contributing to our view of the structure of space-time and the essence of the universe.

The interpretations of quantum mechanics offer various perspectives on the nature of quantum reality. While quantum theory is widely accepted, there are different interpretations regarding the meaning of the mathematical equations and physical phenomena. The Copenhagen interpretation, perhaps the most well-known, focuses on the probabilities associated with quantum measurements and the inherent uncertainty of the quantum world. According to this interpretation, quantum reality is fundamentally probabilistic, with measurement being a distinct act separate from the natural physical process.

On the other hand, the many-worlds interpretation suggests that multiple universes arise from each quantum event, where all quantum possibilities are realized in parallel universes. This transforms measurement into a mere observation of the universe we experience. Beyond these, other lesser-known interpretations, such as the de Broglie-Bohm interpretation and quantum information interpretation, also provide distinct approaches to understanding the quantum nature of reality. Each interpretation brings its own implications and criticisms, and the choice between them may depend on the researcher's personal preference or specific application.

The interpretations of quantum mechanics remain an active field of research and debate, inciting new ideas and discoveries about the essence of the physical world. Quantum mechanics, a crucial foundation, describes the behavior of matter and energy at diminutive scales, such as atoms and subatomic particles. Its most prominent feature is the probabilistic nature of events in the quantum world. Unlike classical physics, where laws precisely determine the future state based on the present state, quantum mechanics offers only probabilities for different measurement outcomes.

This probabilistic nature manifests in the famous double-slit experiment, where individual particles exhibit an interference pattern as if they were waves, while their exact position on the detector remains unpredictable and expressed in terms of probabilities. Although this feature may seem peculiar, it has been repeatedly confirmed by experiments and represents one of the fundamental principles of quantum mechanics, with profound implications for our understanding of reality and scientific observation.

Wave-particle duality and the uncertainty principle are fundamental concepts in quantum mechanics, but they address different aspects of the behavior of subatomic particles. Let's explore each of them:

Wave-Particle Duality

Wave-particle duality refers to the property of certain quantum entities, such as electrons and photons, to exhibit characteristics of both particles and waves. Depending on the experiment conducted, these particles can behave in distinct ways:
- **Wave Behavior:** In experiments such as the famous double-slit experiment, when a source of particles (such as electrons or photons) is directed through two slits, they create an interference pattern typical of waves. This indicates that the particles are behaving as waves, interfering with one another.
- **Particle Behavior:** Conversely, when an attempt is made to measure the position of a particle, it behaves like a particle, occupying a specific location in space. In this case, the measurement result reveals the particle at a defined point, without the characteristic of interference.

Thus, wave-particle duality demonstrates that quantum entities do not fit neatly into the classical categories of "wave" or "particle," but possess a hybrid nature that depends on the context of observation.

Uncertainty Principle

The uncertainty principle, formulated by Werner Heisenberg, is a concept that describes fundamental limits on the precision with which certain pairs of physical properties of a particle can be measured simultaneously.

The most well-known example is the relationship between the position and momentum (quantity of motion) of a particle. The principle states that:
- The more precisely we try to measure the position of a particle, the less precisely we can know its momentum, and vice versa.

This principle is not na experimental limitation, but na intrinsic feature of quantum nature. It reflects the idea that quantum reality is fundamentally probabilistic and that there is a limit to what we can know about the state of a particle.

Summary of Differences
- **Scope:** Wave-particle duality refers to the nature of quantum particles that can behave as waves or particles, while the uncertainty principle deals with the limitations in the simultaneous measurement of physical properties such as position and momentum.
- **Implications:** Wave-particle duality reveals the complexity of the description of quantum particles, whereas the uncertainty principle highlights the probabilistic nature and limits of precision in measurements.

Both concepts are fundamental to understanding the behavior of particles in the quantum world and challenge our intuition about how the universe operates at subatomic scales. The double-slit experiment is one of the most emblematic demonstrations of wave-particle duality in quantum mechanics. It illustrates how quantum particles, such as electrons or photons, can exhibit both wave-like and particle-like characteristics depending on how they are observed. Let's explore how this experiment works and what it reveals about duality.

Experiment Setup
1. **Particle Source:** The experiment begins with a source that emits individual particles, such as electrons or photons. These particles are fired towards a barrier that has two narrow slits.
2. **Barrier with Two Slits:** The barrier has two parallel slits, and the particles can pass through either one.
3. **Detector:** Behind the barrier, there is a detector (such as a

photographic plate or na electronic detector) that records where the particles impact after passing through the slits.

Observed Results

Wave Behavior
- **When the Slits Are Open:** If only one slit is open, the particles behave as expected, creating na impact pattern on the detector corresponding to that single slit. However, when both slits are open, na interference pattern emerges on the detector. This pattern is characteristic of waves, where the waves passing through each slit interfere with one another.
- **Interference:** The interference pattern, showing bands of light and darkness, suggests that each particle is behaving like a wave, passing through both slits simultaneously and interfering with itself.

Particle Behavior
- **Measuring the Trajectory:** If we attempt to measure which slit each particle passes through, the interference pattern disappears. Instead, the particles behave like classical particles, resulting in two patterns corresponding to the two slits, as if each particle had passed through only one slit.
- **Collapse of the Wave Function:** The act of measuring or observing the particle's trajectory seems to "collapse" its wave function, which represents a superposition of states, into a defined state where the particle is clearly associated with one of the slits.

Conclusion
The double-slit experiment powerfully demonstrates wave-particle duality:
- **When We Do Not Measure:** The particles exhibit wave properties, creating na interference pattern that suggests they are behaving like waves passing through both slits at the same time.
- **When We Measure:** The particles behave like classical particles, passing through a single slit and creating a pattern that reflects this trajectory.

This experiment illustrates how the behavior of quantum particles can change based on how they are observed, challenging our intuitions about the nature of reality. It is one of the cornerstones of quantum mechanics and highlights the complexity and richness of the quantum world.

As we conclude our exploration of the quantum world, it is fascinating to reflect on the profound implications of this domain in our lives and how we perceive reality. What we have learned so far is not merely a collection of complex physical concepts; it is a new way of understanding the universe and our place within it.

We live in a world where classical physics, with its deterministic laws, has shaped our understanding for centuries. However, as we venture into the quantum realm, we are confronted with a radically different landscape where uncertainty and probability play central roles. This transition can be disconcerting, yet it is deeply enriching. After all, quantum mechanics not only challenges our traditional notions of cause and effect but also offers us a new language to describe the complexity of nature.

When we talk about wave-particle duality, we are reminded that the particles that make up everything around us—from the stars in the sky to the air we breathe—do not behave in simple or predictable ways. They have a dual nature, exhibiting properties of both particles and waves depending on how they are observed. This revelation forces us to question the very essence of reality: does observation shape our experience of the world? If so, what responsibility do we have in our interaction with reality?

The double-slit experiment, with its beauty and complexity, brilliantly illustrates this idea. By observing a quantum phenomenon, we are not only witnessing a physical event but also actively participating in the construction of reality. This interactivity is a powerful reminder that, in many ways, we are co-creators of our experiences. This notion can be liberating, as it invites us to recognize the role we play in shaping our destiny, both individually and collectively.

Furthermore, as we introduce String Theory and its additional dimensions, we begin to glimpse the possibility that reality is much more complex than we can imagine. The idea that there are hidden dimensions beyond the three we perceive suggests that our understanding of the universe is still incomplete. This leads us to reflect on the humility required in the pursuit of knowledge. We are merely beginning to scratch the surface of universal truths.

Quantum mechanics also invites us to reconsider the nature of time and space. Quantum interactions challenge our linear understanding of time, proposing that past, present, and future events may be interconnected in ways we do not yet fully comprehend.

This idea resonates with many philosophical and spiritual traditions that speak of the interconnectedness of all things. After all, if everything is intertwined in a vast quantum fabric, how does this reflect in our daily lives? How can we cultivate a greater awareness of this interconnectedness in our actions and decisions?

As we explore interpretations of quantum mechanics, such as the Copenhagen interpretation and the many-worlds interpretation, we are reminded that science is not merely a quest for answers but also a journey of inquiry. Each interpretation brings new questions and challenges, and the diversity of perspectives in quantum physics reflects the complexity of reality itself. Ultimately, this teaches us that knowledge is a dynamic and evolving process. The pursuit of truth is as much about the questions we ask as it is about the answers we find.

The uncertainty principle, which tells us that we cannot simultaneously know the position and momentum of a particle with absolute precision, serves as a powerful metaphor for life. Often, we find ourselves in situations where we desire certainties, but reality presents us with a mosaic of possibilities. This uncertainty can be uncomfortable, but it is also what makes life rich and full of potential. It is na invitation to embrace ambiguity and complexity rather than seek simple answers to intricate questions.

As we delve deeper into technology, such as quantum computing, we are encouraged to envision a future where computational capabilities expand exponentially, allowing us to solve problems that once seemed impossible. This innovation is not just a technical achievement; it represents a new way of thinking about information and processing. As these technologies evolve, it is crucial that we maintain na ethical dialogue about how to use them for the benefit of humanity. Science, as exciting as it is, must always be accompanied by reflection on its social and moral implications.

Finally, as we reflect on what it means to live in a quantum world, we are led to consider the beauty and fragility of life. Each moment is na intersection of possibilities, a dance of probabilities unfolding. Just like quantum particles, we too are complex beings, shaped by our experiences, interactions, and choices. Understanding the quantum world offers us a new lens through which we can view our lives—not as a linear sequence of events, but as a rich and interconnected tapestry.

Chapter 6: Quantum Entanglement

In this chapter, we will delve into the intriguing universe of quantum entanglement, a peculiar phenomenon of quantum mechanics. Entanglement occurs when two or more particles come into a unified quantum state, such that their characteristics become inseparable, even when separated by vast distances. Initially proposed by Einstein, Podolsky, and Rosen in 1935, the concept of quantum entanglement was only experimentally verified many years later.

As a cornerstone of quantum physics, this phenomenon challenges our understanding of physical reality. Entangled particles seem to maintain an instantaneous connection, regardless of the distance separating them, which brings significant technological and scientific implications, making it one of the most explored and debated topics in contemporary physics.

Quantum entanglement describes how two particles can be so connected that their properties become inseparable. For example, when two particles, such as electrons or photons, emerge simultaneously in an entangled quantum state, their quantum properties, such as spin or polarization, are correlated in such a way that it is impossible to describe one without considering the other.

This connection is instantaneous, transcending the distance between them, and is often referred to as "spooky action at a distance."

In addition to being fascinating, quantum entanglement is fundamental to quantum technology. It plays a crucial role in quantum cryptography, where information is transmitted using pairs of entangled particles to ensure secure communication. Any attempt at eavesdropping or interception would result in a change in the entangled state, making interception immediately detectable.

In quantum computing, entanglement is essential for the functioning of qubits, the fundamental units of information in a quantum computer. These qubits can exist in multiple states simultaneously, and entanglement allows them to interact in ways that are not possible in classical computing.

Quantum entanglement also raises profound philosophical questions, challenging our conventional view of the world, where distant objects have independent properties and interact through local influences. This new perspective suggests that the world may be intertwined in a much more intrinsic way than we imagine, revealing the complexity and interconnectedness of reality.

One of the most intriguing aspects of quantum entanglement is quantum teleportation. This phenomenon occurs when two quantum particles become entangled, and after measuring one of them, the other is instantaneously measured, regardless of the distance that separates them. In the quantum realm, particles do not have defined values before measurement, and the action on one affects the other, even when they are separated.

Quantum teleportation is a complex process that involves creating a pair of entangled particles, measuring one of them, and transferring the resulting information to a distant third particle. Although quantum teleportation does not allow for the transfer of physical objects, it plays a vital role in quantum communication and quantum computing. In quantum communication, teleportation is used to transmit information securely and with encryption. Any attempt at interception is promptly detected, as measuring an entangled particle alters the state of the other, compromising the original information.

In quantum computing, teleportation is one of the fundamental operations that enable the construction of more complex quantum circuits and the execution of algorithms that surpass the capabilities of classical computers. In summary, quantum teleportation challenges our understanding of reality and has the potential to revolutionize communication and computing in the future.

Quantum cryptography, a growing field, utilizes quantum mechanics to ensure greater security in communication between two points. This security is feasible thanks to the unique properties of quantum systems, such as entanglement and Heisenberg's uncertainty principle. Unlike classical cryptography, which relies on the complexity of algorithms, quantum cryptography uses quantum mechanics to secure communication.

This is achieved through qubits, which are quantum bits capable of entanglement. When two qubits are entangled, changes in one affect the other, regardless of distance. This property allows for the generation of secure cryptographic keys. The key is created from measuring the entangled states and is shared between the communicating parties, ensuring immediate detection of any attempt at interception by a third party, which would interrupt the communication.

Although still in its early stages, quantum cryptography already finds applications in sectors such as financial and government institutions. The security provided by this approach could have significant implications for national security, data privacy, and e-commerce.

The phenomenon that occurs when two or more particles come into a unified quantum state is known as quantum entanglement. In this state, the particles become interdependent to such an extent that the characteristics of one particle are correlated with the characteristics of the other, regardless of the distance that separates them.

In quantum entanglement, the quantum properties of the particles, such as spin, polarization, or momentum, are linked in such a way that measuring one property in one of the particles instantaneously affects the other. This means that even if the particles are separated by great distances, the information about their properties remains interconnected.

This phenomenon was first suggested by Einstein, Podolsky, and Rosen in 1935 in a famous paper questioning the completeness of quantum mechanics. Since then, quantum entanglement has been experimentally proven in various situations and is considered one of the most enigmatic and fascinating aspects of quantum physics. Beyond its conceptual challenges to our understanding of reality, quantum entanglement has significant practical applications in fields such as quantum cryptography, quantum computing, and quantum teleportation.

As we reach the end of this chapter on quantum entanglement, I invite you to reflect on the profound implications this phenomenon has not only in physics but also in our understanding of the world and ourselves. Quantum entanglement teaches us that, at a fundamental level, everything is interconnected. This idea is as fascinating as it is challenging, as it forces us to reconsider our perceptions of separation and individuality.

Imagine for a moment the beauty of two particles that, even when separated by vast distances, remain inextricably linked. This instantaneous connection transcends space and time, challenging classical notions of how we interact with the world around us. Quantum entanglement is not just a physical concept; it is a powerful metaphor that can inspire us to view our relationships and interactions differently.

Just as entangled particles, we too are part of a broader fabric of connections, where every action and every choice resonates beyond ourselves.

Consider how this idea of interconnectedness can manifest in our daily lives. Our decisions, emotions, and experiences impact not only ourselves but also those around us. In a world where we often feel isolated or disconnected, quantum entanglement reminds us that, at a fundamental level, we are all intertwined. This reflection can encourage us to cultivate empathy, compassion, and understanding in our interactions, recognizing that, just like quantum particles, our lives are interconnected in ways we often cannot see.

Furthermore, as we consider the practical applications of quantum entanglement, such as in quantum cryptography and quantum computing, we are led to envision a future where information security and communication become more robust and reliable. Emerging technologies offer us a glimpse into a world where we can overcome current limitations and explore new possibilities. As science advances, we are challenged to remain curious and open to innovations that can shape our future.

Chapter 7: Quantum Mechanics and Consciousness

In this chapter, we will delve into a complex and controversial topic in quantum physics: the intersection between quantum mechanics and consciousness. Researchers have been exploring this relationship for decades, generating various theories and hypotheses about the possible connection between these two fields. We will examine some of these theories and reflect on their potential implications.

The relationship between quantum mechanics and consciousness has been a subject of exploration since the 1930s when Werner Heisenberg suggested that the observation of a quantum system could impact the system itself. This idea was refined by other physicists, such as Niels Bohr and Eugene Wigner.

One of the most discussed theories is the wave function collapse hypothesis. According to this theory, consciousness could cause the collapse of a quantum system's wave function, transforming the probability of a specific state into certainty. Another approach suggests that consciousness may be related to quantum non-locality, a phenomenon that occurs when entangled particles have their properties mutually influenced, regardless of the distance separating them. Some researchers propose that consciousness might influence this non-locality and, consequently, the behavior of particles.

Additionally, there are theories that connect consciousness to quantum superposition, where a particle can exist in multiple states simultaneously. Some interpretations suggest that consciousness could select a specific state from the quantum superposition, impacting the outcome of the measurement.

Although these theories linking quantum mechanics to consciousness are controversial and lack consensus, it is undeniable that quantum mechanics has challenged our traditional notions of reality. The interrelation between quantum mechanics and consciousness remains a debated topic in the scientific and philosophical communities. Since the discovery of quantum mechanics, scientists and philosophers have hypothesized that the probabilistic and non-deterministic nature of the quantum world could have implications for our understanding of consciousness.

One of the first concepts to emerge was Schrödinger's "cat," proposed by Erwin Schrödinger in 1935. This thought experiment questions how a macroscopic object, such as a cat, could exist in a quantum state of superposition. From this, the hypothesis arose that the observer's consciousness might be crucial for the wave function collapse and for determining the state of the observed object.

The Copenhagen interpretation, by Niels Bohr and Werner Heisenberg, also addresses this relationship. This theory suggests that measuring a quantum particle "collapses" its wave function into a particular and irreversible state, implying that the observer plays a vital role in the measurement process, possibly through consciousness.

Roger Penrose and Stuart Hameroff venture into the idea that consciousness arises from quantum processes occurring in the brain. They suggest that the brain's microtubules function as quantum computers, generating superposition states that relate to conscious experience. However, many critics contest these ideas, arguing that consciousness can be understood in purely biological and neurological terms.

In summary, the connection between quantum mechanics and consciousness is a controversial and constantly debated topic in the scientific and philosophical community. While some theories suggest a link, others argue that consciousness can be explained by biological and neurological processes. The theory of quantum consciousness explores this possible relationship, suggesting that consciousness and the quantum world are somehow interconnected, where observation and quantum measurement may be influenced by consciousness.

A central idea in this theory is that observation is not a passive event but rather a dynamic interaction between the observer and the observed system. This interaction can influence measurement and, consequently, impact quantum reality. Another approach proposes that consciousness itself is a quantum phenomenon, emerging from quantum interactions in the brain. Various theories attempt to explain the connection between consciousness and quantum mechanics, including the wave function collapse theory, many-worlds interpretation, and quantum entanglement theory. Each of these brings specific implications for the relationship between consciousness and the quantum world.

Despite the controversies and uncertainties, the theory of quantum consciousness is gaining increasing attention. The field is dynamic and evolving, offering potential for a new understanding of the mind, the brain, and reality itself.

The wave function collapse hypothesis is a central concept in quantum mechanics that describes how a quantum system changes from a state of superposition to a defined state as a result of measurement or observation. To understand this hypothesis, it is important to comprehend some fundamental concepts of quantum mechanics.

Wave Function

In quantum mechanics, the wave function is a mathematical description of the state of a particle or quantum system. This function contains all the information about the properties of the system, such as its position, momentum, and other attributes. However, before measurement, the wave function generally represents a superposition of multiple possible states, meaning that the particle can exist simultaneously in different states.

Superposition

Superposition is a fundamental principle of quantum mechanics that allows a particle to be in several states at the same time. For example, na electron can be in multiple different positions or have different spin values simultaneously. This situation continues until a measurement is performed.

Collapse of the Wave Function

When a measurement is made on a quantum system, the wave function collapse hypothesis suggests that the wave function "collapses" from its superposition of states into a defined state. This process is instantaneous and results in the observation of a specific outcome. For example, when measuring the position of a particle, the wave function describing the particle ceases to represent multiple positions and collapses into a single observed position.

Implications of the Hypothesis

This hypothesis has several important implications:

1. **Probabilistic Nature:** The result of a measurement is inherently probabilistic. Before measurement, the wave function provides the probabilities of finding the particle in different states. The collapse of the wave function transforms these probabilities into a concrete outcome.

2. **Role of the Observer:** The hypothesis implies that observation or measurement plays a crucial role in determining the state of a quantum system. This raises philosophical questions about the nature of reality and the observer's role in the quantum universe.

3. **Conceptual Challenges:** The collapse of the wave function is a concept that generates debates among physicists and philosophers. Different interpretations of quantum mechanics, such as the Copenhagen interpretation, many-worlds interpretation, and objective collapse theory, offer varying perspectives on what truly happens during the collapse.

The series "Constellation" is a production that explores the intersections between quantum physics and consciousness, taking viewers on na intriguing journey through some of the most complex and fascinating theories in science. Throughout its episodes, the series addresses not only the fundamental principles of quantum mechanics but also the philosophical and existential questions that arise when we consider the relationship between human consciousness and the quantum universe.

A Journey of Discovery

"Constellation" proposes a deep reflection on the role that consciousness plays in shaping reality. The series presents theories that challenge our traditional notions of what it means to observe and interact with the world. One of the most provocative aspects discussed is the wave function collapse hypothesis, which suggests that the act of observation is not a passive event but rather na active process that can influence the state of the quantum system. This idea confronts us with the possibility that reality is not simply a fixed structure but something shaped by our perception and consciousness.

Throughout the series, viewers are invited to consider how quantum physics can impact our understanding of ourselves and our place in the universe. The series addresses wave-particle duality, superposition, and quantum entanglement, each of these concepts presented in na accessible and engaging manner. It provides na opportunity to explore how these quantum phenomena may have implications not only in technology but also in fields such as philosophy, spirituality, and psychology.

Reflections on Consciousness

One of the central themes of "Constellation" is the nature of consciousness. The series prompts us to reflect on what it means to be conscious in a world where quantum rules seem to contradict our everyday experience. The notion that consciousness may play na active role in the collapse of the wave function suggests that we are more than mere passive observers; we are active participants in the creation of reality. This can be na empowering perspective, suggesting that our intentions and perceptions have a real impact on the world around us.

Moreover, "Constellation" challenges us to consider the interconnectedness of all beings. By discussing quantum entanglement, the series touches on the idea that everything in the universe is somehow linked, even when it seems distant. This notion of interconnectedness may resonate strongly with many spiritual and philosophical traditions that emphasize the unity and interdependence of all things. The series invites us to expand our view of how we relate to one another and the natural world, recognizing that our actions and thoughts have na echo that resonates beyond ourselves.

Ethical and Practical Implications

"Constellation" also raises ethical questions about the use of quantum knowledge and its technological implications. As we move toward innovations such as quantum computing and quantum cryptography, we are confronted with the responsibility to use these technologies ethically and consciously. The series suggests that as we expand our understanding of the quantum universe, we must consider the impact our actions will have on society and the environment. This ethical reflection is especially relevant at a time when technology advances rapidly and presents significant challenges.

The Role of Interpretation

Another interesting point addressed in the series is the diversity of interpretations of quantum mechanics. "Constellation" reminds us that, just as in science, our perceptions and interpretations of reality are shaped by our cultural context, personal experience, and worldview. The series presents different interpretations of quantum mechanics, such as the Copenhagen interpretation and the many-worlds interpretation, stimulating viewers to consider how these theories may influence our understanding of reality and consciousness.

Conclusion: Na Invitation to Reflection

At the end of "Constellation," we are left with a sense of wonder and curiosity about the mysteries of the universe and the nature of consciousness. The series does not provide definitive answers but instead invites us to explore these questions openly and reflectively. This approach is crucial, as it reminds us that the pursuit of knowledge is na ongoing journey, filled with questions and discoveries.

We have come to the end of this chapter that explores the intriguing relationship between quantum mechanics and consciousness, a theme that leads us to question the boundaries of what we know about reality and ourselves. Throughout this journey, we have reflected on how quantum physics is not just a set of laws governing the behavior of particles but also na invitation for introspection and reconsideration of our role as observers in this vast universe.

The idea that consciousness can influence the quantum world is undoubtedly one of the most fascinating and provocative. The wave function collapse hypothesis, for example, suggests that by observing, we are not merely recording a state but actively participating in the creation of reality. This notion leads us to contemplate the depth of our interactions with the world around us. Each of us, with our perceptions and intentions, can shape reality in ways that go beyond what classical physics might lead us to believe.

It is intriguing to think about how this quantum perspective might apply to our daily lives. If we are indeed interconnected at a quantum level, this could inspire us to cultivate a greater awareness of our actions and their implications. The small choices we make—from how we communicate with others to how we care for our environment—can resonate in ways we cannot see but that certainly have na impact.

When we talk about Schrödinger's "cat" and superposition, we are reminded that reality is not always as binary as it seems. Often, we live in a space of uncertainty, where multiple possibilities coexist. This understanding can be liberating, as it encourages us to embrace the ambiguity and complexity of life. Instead of seeking definitive answers, we can learn to appreciate the process of exploration and discovery.

The series "Constellation," which I mentioned earlier, is na excellent reflection of how these ideas can be explored in na accessible and engaging manner. It invites us to dive into the nuances of quantum mechanics while also challenging us to think critically about the role of consciousness. Watching this series can be a revealing experience, opening our minds to new possibilities and expanding our understanding of the connection between science and spirituality.

As we delve deeper into theories linking quantum mechanics to consciousness, we also face skepticism. It is natural for science to seek solid evidence before embracing new ideas. However, the beauty of scientific inquiry lies in its ability to evolve. What may seem like a bold speculation today could one day be recognized as a fundamental truth. This is the essence of progress: to question, investigate, and eventually discover.

Another point to consider is the ethical impact of discoveries at the intersection of quantum mechanics and consciousness. As we develop technologies based on quantum physics, such as quantum computing and quantum cryptography, we are reminded that we have a responsibility regarding the use of this knowledge. What we do with the tools we build can shape the future of humanity. Therefore, it is vital that we approach these innovations with na ethical mindset, aware of their implications.

Finally, I invite you to reflect on how the ideas discussed in this chapter may relate to your own life. How do you perceive your own consciousness? In what ways does your perception of the world influence your interactions with others? What does it mean for you to live in a universe where everything is interconnected?

The journey through quantum mechanics and its relationship with consciousness is na ongoing exploration, full of questions and possibilities. We do not have all the answers, but the pursuit of them is what makes this journey so rich and meaningful. As we move forward, may we remain curious, open, and willing to question what we know, allowing new ideas to flourish and guide us in our understanding of the universe.

Chapter 8: Quantum Physics and Spirituality

In Chapter 8, the fabric of quantum physics intertwines with spirituality, a theme that has sparked profound reflections since the revelation of quantum mechanics. Researchers and philosophers have speculated whether this theory can shed light on non-physical phenomena such as consciousness and spirituality, while also questioning the very nature of reality. Although the connection between quantum physics and spirituality remains a contentious point, a growing body of evidence suggests that quantum theory may help us understand the nature of consciousness, universal interconnectedness, and the fundamental essence of reality.

In this chapter, we will delve into the possible intersections between quantum physics and spirituality. We will deliberate on how quantum theory can illuminate spiritual phenomena such as meditation, energy healing, and intuition. We will also explore how these ideas may influence our self-perception, our views of others, and our understanding of the world around us.

We will address the criticisms and challenges this relationship faces and examine the implications it may have for our perception of reality and the role of consciousness in the cosmos. The intersection of quantum physics and spirituality is a crucial meeting point, emerging as one of the peaks of this book. We will begin by investigating how quantum mechanics can shed light on certain spiritual experiences. For instance, quantum entanglement may serve as na analogy for the universal connection present in many spiritual traditions. Similarly, the probabilistic nature of quantum mechanics suggests that perception shapes reality, a concept that resonates with various spiritual philosophies.

Na essential approach is to explore the nature of consciousness. While many argue that consciousness is na emergent property of the brain, quantum mechanics suggests it may be more intrinsic, present throughout the cosmos. This invites us to a deeper understanding of the mind and the role of consciousness in the universe.

Additionally, the idea that reality is forged by perception may lead to a more profound understanding of reality itself. Many spiritual traditions assert that reality is illusory, and quantum mechanics may illuminate the validity of this claim. We will also investigate how quantum physics can help us understand spiritual phenomena such as spiritual healing. Quantum mechanics suggests that perception can influence reality, providing a possible explanation for the effectiveness of spiritual healing.

In conclusion, this chapter will be a journey to explore the possible links between quantum physics and spirituality, and how this convergence can lead us to a deeper understanding of the nature of the universe and our own existence.

Understanding Spirituality in Light of Quantum Physics

Quantum physics, widely debated and applied in various fields, still generates controversy when related to spirituality. Many scholars and thinkers see quantum physics as a new perspective for understanding spirituality and concepts such as energy, consciousness, connection, and the universe. Quantum theory offers us the opportunity to interpret the world in terms of interconnectedness and relationship, as opposed to isolated and autonomous entities. Similarly, many spiritual traditions advocate for the existence of a universal connection and the interdependence of all things.

Quantum physics also sheds new light on time and space, where reality is influenced by observation, not presenting itself as something fixed or objective. This idea can be associated with the spiritual notion that reality is a subjective construction of the mind, and that our consciousness has the power to shape and influence our experiences.

Another vital connection between quantum physics and spirituality lies in the concept of energy. From a quantum perspective, energy is seen as the foundation of all matter, expressed through oscillations and vibrations. In spiritual traditions, energy is considered a life force that permeates everything, influencing our physical and emotional health. It is argued that quantum physics can illuminate spiritual experiences such as meditation and the connection to something greater than ourselves. The assumption is that quantum physics can reveal how our mind and consciousness are intrinsically linked to the essential nature of the universe.

However, it is essential to note that many scientists and scholars remain skeptical about this intersection between quantum physics and spirituality. Some assert that these two realms are entirely distinct and that the attempt to intertwine them is misguided. Others argue that it is possible to embrace spiritual understanding without resorting to quantum physics. Despite the varied perspectives, it is undeniable that both quantum physics and spirituality challenge our conceptions of reality and the nature of the universe. Both seek to decipher the world in a deeper and more meaningful way, and the connection between them continues to be a topic of study and debate.

The Unity of the Universe

Understanding the unity of the universe is a pillar in both quantum physics and spirituality. While science strives to elucidate the physical nature of the cosmos, spirituality focuses on understanding consciousness and existence. Despite appearances, quantum physics and spirituality may share a converging perspective on the nature of reality. Quantum physics reveals that reality is not as solid and predictable as once thought. At the quantum level, subatomic particles exist in states of superposition, being able to occupy more than one place or state simultaneously. Furthermore, quantum mechanics demonstrates that observation shapes the outcome of a measurement, suggesting that consciousness participates in the creation of reality.

On the other hand, spirituality teaches that reality is not only physical but also energetic and spiritual. The belief in a universal consciousness that interconnects all beings and mutually influences them is a common premise in many spiritual traditions. By understanding that reality is not only physical but also energetic and spiritual, we can see how quantum physics and spirituality may converge in a view of the nature of reality. Both perspectives point to a unified and intertwined reality where consciousness and matter are intricately connected.

In this context, understanding the unity of the universe can lead us to a deeper and more meaningful perception of reality. Science helps us unravel the physical nature of the cosmos, while spirituality sheds light on the essence of consciousness and existence. Together, these approaches can provide us with a more comprehensive and holistic understanding of the nature of reality.

Quantum Interconnection of Life

The quantum interconnection of life emerges as a theme that posits that all things in the universe are intrinsically united, and quantum physics can help elucidate this interconnection. According to quantum physics, subatomic particles are intertwined through intricate quantum interactions known as quantum entanglement. This quantum interconnection can echo at all levels of life, from subatomic particles to living beings.

One implication of this quantum entanglement is that all things are interconnected, transcending traditional disciplinary divisions. For instance, biology can be viewed as a unfolding of quantum physics, and the latter, in turn, as a derivation of biology. This is due to the presence of quantum interactions at all levels of life, from molecules to entire organisms.

Another aspect of quantum entanglement is that the separation between observer and observed may be na illusion. In quantum mechanics, measuring a quantum particle can instantaneously alter the state of another particle, regardless of the distance separating them. This phenomenon suggests that the division between things may be na illusion fostered by our limited perceptions. Some scholars suggest that quantum entanglement may explain phenomena such as telepathy, clairvoyance, and other forms of non-localized communication. From this perspective, such occurrences may be consequences of the quantum interconnection that allows for instantaneous communication between individuals. However, this viewpoint raises controversies and remains under scientific scrutiny and investigation.

The relationship between quantum mechanics and consciousness, as discussed in the text, suggests that there are significant interconnections between these two fields that can shed light on the nature of reality and human experience. Here are the main points highlighted in the text:

1. **Intersection of Quantum Physics and Spirituality:** Quantum mechanics is seen as a possible key to understanding non-physical phenomena such as consciousness and spirituality. This connection is considered a point of contention, but there is a growing body of evidence suggesting that quantum theory may help in comprehending the nature of consciousness and universal interconnectedness.

2. **Universal Connection:** Quantum entanglement is used as an analogy for the universal connection present in many spiritual traditions. This implies that just as entangled particles are interconnected, humans and all forms of life are also deeply connected.

3. **Nature of Consciousness:** The text explores the idea that consciousness is not merely an emergent property of the brain but may be an intrinsic characteristic that permeates the entire cosmos. This perspective invites a broader understanding of the role of consciousness in the universe.

4. **Reality as a Subjective Construction:** Quantum mechanics suggests that observation and perception shape reality, aligning with many spiritual traditions that assert reality is a subjective construction of the mind. This idea challenges the notion of an objective and fixed reality.

5. **Energy and Spiritual Healing:** The concept of energy is central both in quantum physics and spirituality. Energy is viewed as the foundation of all matter in quantum mechanics and simultaneously as a vital force in many spiritual traditions. This intersection may offer explanations for phenomena such as spiritual healing, where perception and intention can influence reality.

6. **Criticism and Skepticism:** While there is a quest for connections between quantum physics and consciousness, there are skeptics who argue that these areas are distinct and that attempting to intertwine them may be a mistake. However, the pursuit of understanding the interconnection between these fields remains a relevant topic of debate.

Conclusion and Reflection on the Intersection of Quantum Physics and Spirituality

As we reach the end of this chapter, it is essential to reflect on the rich tapestry that forms when we intertwine quantum physics and spirituality.

The journey we have undertaken has led us to consider how these two seemingly distinct disciplines can, in fact, provide profound insights into the nature of reality and the role of consciousness in our lives.

Challenging Our Perceptions

Quantum mechanics, with its paradoxical truths and surprising phenomena, challenges us to re-examine our traditional perceptions of the world. The idea that observation shapes reality is a powerful notion that invites us to reconsider not only how we view the universe but also how we interact with it. Each of us, as observers, plays an active role in creating our own reality, shaping the experiences we live through our intentions, perceptions, and actions.

This perspective is similar to the spiritual view that teaches us that reality is a subjective construction of the mind. When we combine these ideas, we begin to realize that we have greater control over our lives than we often admit. By becoming aware of our beliefs and emotions, we can transform them into tools to shape our experiences in a way that aligns more closely with our deepest desires and values.

The Universal Connection

One of the central themes that emerge from this intersection is the universal connection that unites all beings. The concept of quantum entanglement reveals to us that, at a fundamental level, we are all interconnected. This idea resonates in many spiritual traditions that speak of the interdependence of all forms of life. By recognizing this connection, we are invited to cultivate a sense of responsibility not only for ourselves but also for the world and the people around us.

This interconnectedness challenges us to think about how our actions impact others and the environment. Every thought, every word, and every act has the potential to reverberate through this network of interconnection. Therefore, it is our responsibility to act consciously, promoting harmony and well-being not only in our lives but also in the collective.

The Nature of Consciousness

The discussion about the nature of consciousness is one of the most intriguing that emerges from this reflection. Quantum mechanics suggests that consciousness may be a fundamental characteristic of the universe, not merely a consequence of neurological processes. This invites us to consider that consciousness could be a force that permeates everything, influencing reality in ways we do not yet fully understand.

This view expands our understanding of who we are and what our role is in the cosmos. If consciousness is an intrinsic property of the universe, this implies that each of us is part of something much larger. This interconnectedness leads us to question how we can live more authentically, aligning our lives with this broader reality.

The Power of Intention

As we explore the relationship between quantum physics and spirituality, we are led to consider the power of intention. The idea that our perception and intention shape reality suggests that we have the ability to create significant changes in our lives and the lives of others. This is particularly relevant in spiritual practices such as meditation, where intention can be used as a powerful tool for personal transformation and healing.

We invite you, the reader, to reflect on your own intentions and how they may be shaping your reality. What are the thoughts and beliefs you nurture? How can you use them to create a life that resonates with your deepest values? The power of intention is a tool that we can all access and utilize to improve our lives and positively impact those around us.

Challenges and Criticism

It is also important to recognize that this relationship between quantum physics and spirituality is not without challenges and criticisms. Many scientists argue that trying to intertwine these two fields can lead to confusion and misunderstandings. This skepticism is valid and necessary in the pursuit of a clearer and more grounded understanding. However, even amidst these criticisms, the quest for connections between science and spirituality remains a relevant and intriguing topic.

Through this journey, we have learned that exploring the unknown is an essential part of advancing knowledge. Science and spirituality can coexist, and by intertwining, they can offer us a richer and more holistic understanding of life. This search is an invitation to curiosity and questioning, which are fundamental to personal and collective growth.

Walking Together

As we conclude this chapter, it is essential to remember that the exploration of quantum physics and spirituality is an ongoing journey. Each of us brings our own experiences, beliefs, and questions to this discussion. By sharing our reflections and learning from one another, we can enrich our understanding of the nature of reality and consciousness.

Chapter 9: Quantum Physics and the Evolution of Human Consciousness

In this ninth chapter, we will embark on na intriguing reflection on the evolution of human consciousness in response to advancements in quantum physics. Since its discovery, quantum mechanics has revolutionized our understanding of reality, challenging long-standing principles and beliefs. We will explore how these discoveries may have played a fundamental role and will continue to influence the transformation of our consciousness, both individually and collectively.

Adopting na interdisciplinary approach, we will examine how advancements in quantum physics can enrich our view of humanity and our interaction with the world around us. The path that quantum physics has opened for the evolution of human consciousness is filled with fascinating possibilities.

One intriguing aspect is the notion that everything in the universe is interconnected, and that each of us is part of na intricate cosmic web. This deep connection is expressed in quantum physics through the principle of non-locality, which shows us that quantum particles can remain linked even when separated by vast distances. This understanding of interconnectivity can serve as a catalyst for the evolution of our consciousness, awakening deeper compassion, empathy, and a heightened sense of social responsibility.

As we internalize the idea that all things are connected, our concerns expand to include not only our own well-being but also the well-being of all beings and our planet. This shift in perspective can inspire actions that promote harmony and care for others and the environment.

Another important aspect through which quantum physics can stimulate the evolution of our consciousness is the crucial role that the observer plays. In quantum mechanics, the simple act of observing a quantum particle can influence the outcome of the measurement, suggesting that our consciousness is na active factor in the creation of reality. This understanding can deepen our perception of the power of the mind and intention in shaping the reality we experience.

As more people become aware of quantum principles, it is likely that the acceptance of practices such as meditation and creative visualization—which work with the mind to influence reality—will expand. These practices can become valuable tools for cultivating a higher consciousness and a positive impact on our lives.

Throughout this chapter, we will explore these ideas more deeply and examine how quantum physics can be a catalyst for the evolution of our consciousness, both individually and collectively. As we reflect on the evolution of human consciousness in light of advancements in quantum physics, we realize that these discoveries have shed new light on reality and consciousness, which are intrinsically intertwined.

The application of these principles can pave the way for significant social and environmental transformations, through the understanding of universal interconnectedness and the active role of our consciousness in shaping the world around us. Quantum physics has the capacity to stimulate this evolution in various ways, one of which is the perception that everything in the universe is interconnected, with each individual being part of a greater whole.

This interconnectivity, grounded in quantum physics, suggests that quantum particles, even when physically separated, can be entangled in a state of interdependence. This perception can inspire the progress of human consciousness, fostering greater compassion, empathy, and social responsibility. By perceiving the interconnectedness of all things, our care extends not only to our own well-being but also to the well-being of others and the environment.

Another means by which quantum physics can drive the evolution of human consciousness is through the role of the observer. In quantum mechanics, observing a quantum particle can influence the outcome of the measurement, suggesting that the observer's consciousness has na active role in reality. This understanding can promote greater awareness of the power of the mind and intention in constructing reality.

As knowledge of quantum physics spreads, the acceptance of practices such as meditation and creative visualization, which use the mind to shape reality, is likely to increase. Lastly, quantum physics can inspire the evolution of human consciousness by challenging entrenched materialist paradigms in science and Western culture, suggesting that the world is not merely a cold, mechanical machine, but a living, interconnected system.

To support these ideas, we can refer to several renowned authors, such as Fritjof Capra in "The Web of Life: A New Scientific Understanding of Living Systems," Amit Goswami in "The Physics of the Soul: Science and Spirituality in the Construction of a New Paradigm," and William A. Tiller in "Science and Human Transformation: Subtle Energies, Intentionality, and Consciousness."

The principle of non-locality in quantum physics refers to the idea that subatomic particles can be interconnected in such a way that measuring or altering the state of one particle can instantaneously affect another particle, regardless of the distance separating them. This phenomenon has been evidenced in experiments, such as those related to quantum entanglement, where pairs of particles can be generated in such a way that their properties are correlated, even when separated by large distances. Thus, a change in one particle is immediately reflected in the other, challenging traditional notions of causality and locality, which assert that na effect must have a nearby cause.

Relation to Human Consciousness

The connection between the principle of non-locality and human consciousness is na intriguing and often debated topic. Some interpretations suggest that, just as quantum particles can be entangled and influence each other non-locally, human consciousness may also be interconnected at a deeper level. Here are some ways this relationship is explored:

1. **Universal Interconnection:** The idea of non-locality suggests that all things in the universe are intrinsically interconnected. This may resonate with many spiritual traditions that emphasize the interconnectedness of all beings. When we consider that consciousness may be part of this network of interconnection, it can promote a deeper understanding of how our actions and intentions affect not only our lives but also the lives of others and the planet.

2. **Power of Intention:** Quantum mechanics suggests that observation and consciousness play active roles in shaping reality. If consciousness can influence the state of a particle, as proposed in quantum mechanics, this can lead to reflection on the power of human intention. The recognition that our thoughts and intentions can have a real impact may inspire practices that seek to align consciousness with the creation of a more positive reality.

3. **Collective Experiences:** The principle of non-locality can also be seen as a parallel to collective experiences of consciousness. When groups of individuals come together in intention or meditation, for example, some believe that this union of consciousness can create na effect that transcends the sum of its parts, similar to quantum entanglement.

4. **Challenges to Materialist Paradigms:** Non-locality challenges traditional materialist views of science, which often consider reality as a purely mechanical system. This new perspective may open space for greater acceptance of spiritual views that recognize the importance of consciousness and energy, suggesting that consciousness is not merely a product of the brain but a force that permeates reality.

Quantum mechanics suggests that consciousness can influence reality through various principles and phenomena that challenge our traditional understanding of the world. Here are some of the main ways this influence is proposed:

1. **The Role of the Observer:** One of the most fundamental concepts in quantum mechanics is the role of the observer. The famous thought experiment of "Schrödinger's cat" illustrates that, until na observation is made, a quantum particle can exist in multiple states simultaneously (superposition). When na observer makes a measurement, the particle "collapses" to a specific state. This idea suggests that mere observation can affect the outcome of na experiment, raising questions about the active role of consciousness in shaping reality.

2. **Collapse of the Wave Function:** The wave function in quantum mechanics describes the probability of finding a particle in a certain state. The collapse of the wave function occurs when a measurement is made, resulting in a defined state. This transition between superposition and a determined state leads to discussions about whether the observer's consciousness is necessary for this collapse. Some theorists argue that consciousness may be a factor that determines the outcome, implying that the mind has na active role in the creation of observed reality.

3. **Quantum Entanglement:** Quantum entanglement is a phenomenon where particles become interconnected such that the state of one particle is directly related to the state of another, regardless of the distance separating them. This phenomenon challenges the notion that interactions can only occur between nearby objects. The idea that consciousness may be interconnected at a quantum level suggests that the mind can have na impact on events occurring at a distance, reinforcing the notion of interconnectedness among all beings.

4. **Intention and Reality Creation:** Some approaches suggest that intention can influence quantum reality. The belief that our thoughts and intentions can shape reality resonates with spiritual and psychological practices, such as creative visualization and meditation. When individuals focus on a desired outcome, this may create a field of influence that could, theoretically, affect the quantum state of events or situations.

5. **Challenges to Materialism:** Quantum mechanics challenges the materialist paradigms that dominate traditional science, proposing that the universe is not a cold, mechanical machine but a living, interconnected system. This new view may open space for the acceptance that consciousness is not merely a byproduct of the brain

but na active force that interacts with reality. This perspective can encourage a more holistic understanding of life, where consciousness and matter are intrinsically linked.

Final Reflections on Quantum Physics and the Evolution of Human Consciousness

As we conclude our exploration of the relationship between quantum physics and the evolution of human consciousness, it is important to pause and reflect on the profound implications these ideas may have in our lives. This chapter is not merely a collection of scientific concepts; it is na invitation for each of us to reexamine how we see the world and our place within it.

The Interconnection of the Cosmos

Quantum physics offers us a new lens through which we can observe reality. The idea that everything in the universe is interconnected, expressed through the principle of non-locality, reminds us that we are part of a vast cosmic web. This interconnection can inspire us to cultivate a sense of responsibility not only for ourselves but also for others and the environment that surrounds us. When we understand that our actions and intentions have na impact that transcends the individual, we are compelled to act more consciously and compassionately.

The Power of Consciousness

Another important point that has emerged is the active role of consciousness in creating our reality. Quantum mechanics suggests that observation and intention can shape the reality around us. This is a powerful reminder that each of us has a significant role in how we perceive and interact with the world. By becoming more aware of our thoughts and actions, we can begin to direct our energy toward creating realities that are more aligned with our values and aspirations.

Personal and Collective Transformation

Quantum physics not only challenges the materialist paradigms that have dominated Western thought, but it also opens space for greater acceptance of spiritual perspectives. This movement toward a more holistic view is na opportunity for each of us to reflect on how we can integrate these principles into our daily lives. Personal and collective transformation begins with the willingness to see beyond divisions and recognize the interdependence of all things.

Practices that Cultivate Consciousness

As we become more aware of interconnection and the power of intention, practicing techniques such as meditation, visualization, and mindfulness can become valuable tools. These practices not only help us calm the mind but also align our consciousness with our higher goals. By dedicating ourselves to cultivating a positive and intentional mindset, we can influence not only our personal reality but also contribute to significant social and environmental changes.

A Call to Action

I invite you, dear reader, to consider how these ideas resonate in your life. What does it mean for you to be part of na interconnected web? How can you use your consciousness and intention to shape a reality that reflects your values? Each of us has the ability to be na agent of change, and consciousness is a powerful tool we can use to promote transformation.

The Future of Human Consciousness

As we move into the future, the evolution of human consciousness is in our hands. As more people become aware of quantum principles and their implications, we can expect a cultural shift that values interconnectedness, compassion, and social responsibility. This transformation will not be instantaneous, but each step we take toward greater understanding and acceptance can have a domino effect.

The Potential of Quantum Consciousness in Human Evolution

The potential of quantum consciousness to drive the evolution of humanity is immense and deserves to be explored and incorporated into our daily lives. As we delve deeper into the influence of quantum physics on the evolution of human consciousness, additional aspects emerge that deserve our consideration. Beyond the interactions already discussed between quantum mechanics and consciousness, it is essential to examine how these concepts manifest in areas such as psychology, spirituality, and creativity.

1. **Quantum Psychology: Exploring the Mind from a New Perspective**

The emerging field of quantum psychology seeks to draw parallels between quantum principles and human mental processes. The uncertainty present in quantum mechanics may resemble the fluid nature of thoughts and emotions. Thus, superposition, which allows particles to exist in multiple states simultaneously, reflects the complexity and multiplicity of perspectives that the human mind experiences.

When we apply these principles to psychology, new ways of understanding decision-making and perception emerge. The idea that a decision may exist in different states before being "observed" represents the exploratory nature of the mind as it considers various options. Moreover, quantum perception suggests that the way we see something can affect its manifestation, highlighting how our subjective interpretation shapes reality.

Emotions can also be viewed as complex quantum interactions. The way we feel and process na emotion has a direct impact on our experience of reality. The interconnection between emotions and responses to external events can be compared to quantum entanglement, where separated particles are intricately correlated.

2. **Quantum Spirituality and Inner Transformation**

The fusion of quantum principles and spirituality seeks to find common ground between these seemingly distinct areas. Quantum spirituality proposes that the underlying reality is interconnected and that consciousness influences this interconnection, resonating with spiritual views of fundamental unity.

In this context, the mind is seen as na active creator of reality. The belief that our thoughts and mental states shape our experience gains a new dimension in light of quantum principles, suggesting that the mind has a more subtle and complex role than we imagine.

The spiritual quest for inner transformation aligns with quantum understanding, demonstrating that consciousness is in constant evolution, reflecting the adaptability and change of the universe.

Creativity and Innovative Thinking

Understanding superposition in quantum physics can inspire new approaches to creativity. The notion that na idea can exist in multiple states simultaneously suggests that exploring different perspectives may lead to more original and innovative solutions. Similarly, non-locality teaches us that creative insights can occur instantaneously, transcending conventional boundaries of the mind and connecting ideas in unexpected ways.

Ethics and Global Responsibility

Quantum interconnectedness can influence our ethical choices and global responsibility. By recognizing that our individual actions reverberate throughout the interconnected system, we are encouraged to make more conscious and compassionate decisions that benefit not only ourselves but also humanity and the planet. This understanding leads us to reflect on quantum ethics, reminding us that our responsibility extends beyond ourselves to all beings and the environment. By making decisions with this awareness, we contribute to a more harmonious and balanced world.

Conclusion: Embracing the Evolution of Quantum Consciousness

As we conclude this chapter, we are invited to reflect on the profound interconnection between quantum physics and the evolution of human consciousness. By exploring the ramifications of quantum principles in various areas of our lives—from psychology to spirituality, from creativity to ethics—we are led to contemplate the vast potential unfolding before us.

Quantum psychology teaches us that our mental processes are complex manifestations of entanglement and superposition, expanding our understanding of the human mind and opening doors to a more holistic approach to our own nature. Quantum spirituality invites us to look beyond the boundaries of matter, perceiving the deep connection between all beings and the cosmos, reminding us that consciousness is the foundation of everything.

By embracing quantum creativity, we discover that the mind can explore multiple possibilities simultaneously, unleashing innovative thoughts that shape the future. The awareness that ideas can exist in various states empowers us to adopt a more expansive and imaginative approach in our creative endeavors.

As we close this chapter, remember that each of us is part of this collective experience. Our journey is shared, and together we can create a more harmonious, interconnected, and conscious future. May we move forward with open hearts and curious minds, ready to explore the infinite possibilities that life offers us.

The Quantum Realm Experienced by Me

My experience of the quantum realm highlights my unique journey through the exploration of psychedelic substances, allowing me to immerse myself in the quantum interconnectedness that permeates all things. During the electronic music festival called Soul Vision, I allowed myself to enter a universe of colors and shapes I had never witnessed, yet found strangely familiar. It was the experience with DMT that defined a radical shift in my perspective. My consciousness expanded, enabling me to glimpse the entirety of the universe in a single vision. In that moment, I experienced a sense of unity with all things and a profound understanding of the inherent interdependence and interconnectedness of everything.

Since then, I have directed my life toward understanding these phenomena and their transformative potential. I have dedicated myself to researching and theorizing about the relationship between quantum physics, spirituality, and consciousness, recognizing how this understanding can lead to significant changes in society and the environment. My experience at the music festival was a catalyst that awakened the importance of the convergence between science and spirituality. It became evident to me that the beauty of the universe and our intrinsic connection to it can be contemplated through various lenses, whether through art, meditation, or science.

From this experience, I internalized that true transformation emerges from self-knowledge and awareness of our interconnectedness with all things. This understanding enables the construction of a more promising future, where science and spirituality unite to foster a deeper perception of the universe and our relationship with it. I recognize that my journey in seeking knowledge and expanding consciousness is continuous and ongoing.

I understand that to connect fully with the universe, it is necessary to be in tune with the perception that the universe and we are inseparable entities, observing the multitude of infinite possibilities. The assimilation of quantum physics and spirituality, combined with my experience at the Soul Vision festival, allowed me to develop my own theory, grounded in the seven keys of mental alchemy and the Quantum Universe. This theory not only guided me to make a positive impact on the world but also provided me with a deeper and more comprehensive connection to the universe. I understand that the pursuit of knowledge and the expansion of consciousness is na unending journey, requiring a mind always open to new perspectives and possibilities. My commitment to sharing my discoveries aims to inspire others to embark on their own journeys of exploration.

Thus, I conclude that the union of quantum physics and spirituality opens doors to renewed perceptions and understandings of the world around us. My search for knowledge and connection with the universe can lead us to unimaginable horizons. Through the unique experiences I lived at the Soul Vision festival, I navigated the boundaries of science and spirituality, using high doses of psychedelics and investigating theories in practice. This trajectory led me to formulate research and theories anchored in the concepts of quantum physics and spirituality.

My work, "The Alchemist of the Future," is a synthesis of my discoveries about the seven keys of mental alchemy and their interconnection with quantum physics, as well as my explorations of the Quantum Universe, in which consciousness is the foundation of reality. The book represents a significant contribution to the understanding of the interconnection between mind, universe, and spirituality. I believe that the conscious and responsible exploration of the depths of the mind can result in impactful discoveries for humanity.

This pioneering approach at the intersection of science and spirituality opens new perspectives for human evolution and the ability to transcend limitations, achieving a higher understanding of life and the universe. This understanding leads me to a deep sense of humility and gratitude. Every moment of our lives represents na opportunity to explore and expand our consciousness, to learn more about who we are and what our place in the universe is. With each choice we make, we are creating new possibilities and realities, not just for ourselves, but also for others.

As we move forward in our journeys, facing challenges and discovering new possibilities, it is important to remember that we are always connected to each other and to the universe as a whole. When we recognize this interconnectedness, we can collaborate to build a better world, grounded in values such as compassion, empathy, and love. Therefore, I invite you to open your mind and heart to the infinite possibilities that the universe has to offer. Be curious, be courageous, be authentic. And never forget that we are one, all part of the same cosmic fabric, experiencing endless possibilities.

It is undeniable to recognize my role in these valuable contributions to the understanding of science and spirituality, and for my courage in exploring the frontiers of the mind and the universe, in search of deeper knowledge about life and consciousness.

With love and affection,
Muriel Fernandes
The Alchemist of the Future

AN CHEMICAL SCIENTIST OF FUTURE YICEMITSELLING ATHOR OF

THE ALCHEMIST OF THE FUTURE TRILOGY

THE CHEMICAL ALCHEMIST   THE FUTURE TRILOGY

www.ingramcontent.com/pod-product-compliance
Lightning Source LLC
Chambersburg PA
CBHW070131230526
45472CB00004B/1508